This Book Belongs

PATIENCE IS BITTER, BUT ITS FRUIT IS SWEET.

~ Jean-Jacques Rousseau

8.5" x 11" (21.59 x 27.94 cm)
Non Perforated
1/10 inch 10x10 Grid
100 Pages

Page	Topic	Page	Topic

Page	Topic	Page	Topic

Page	Topic	Page	Topic

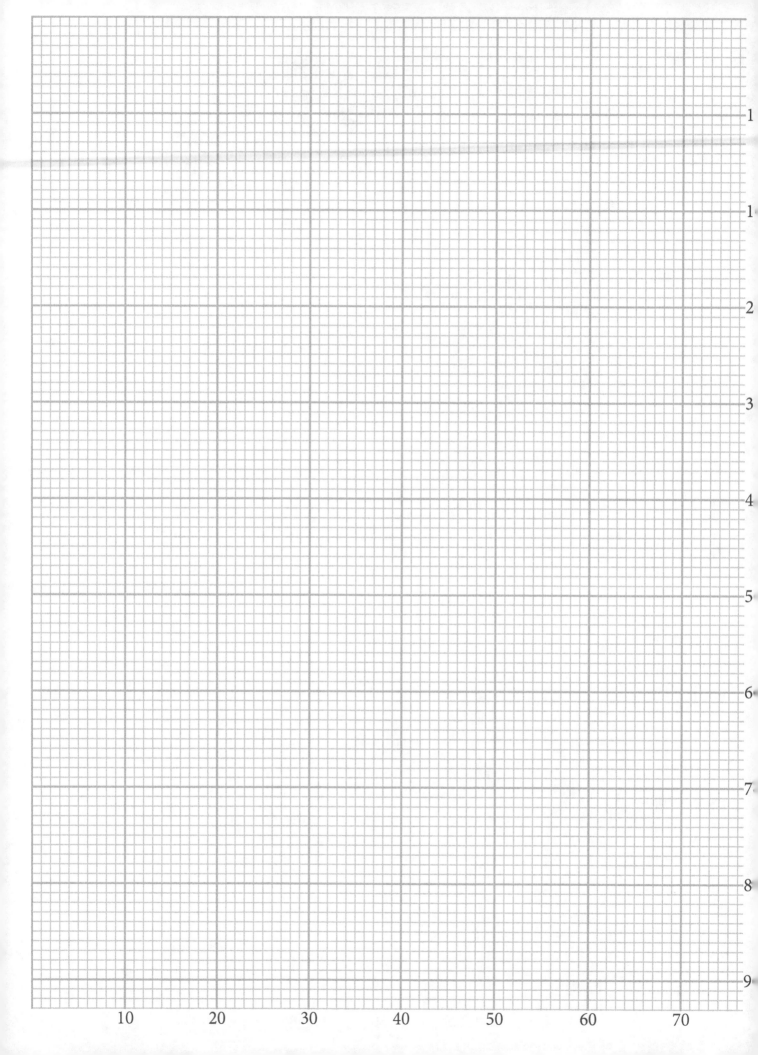

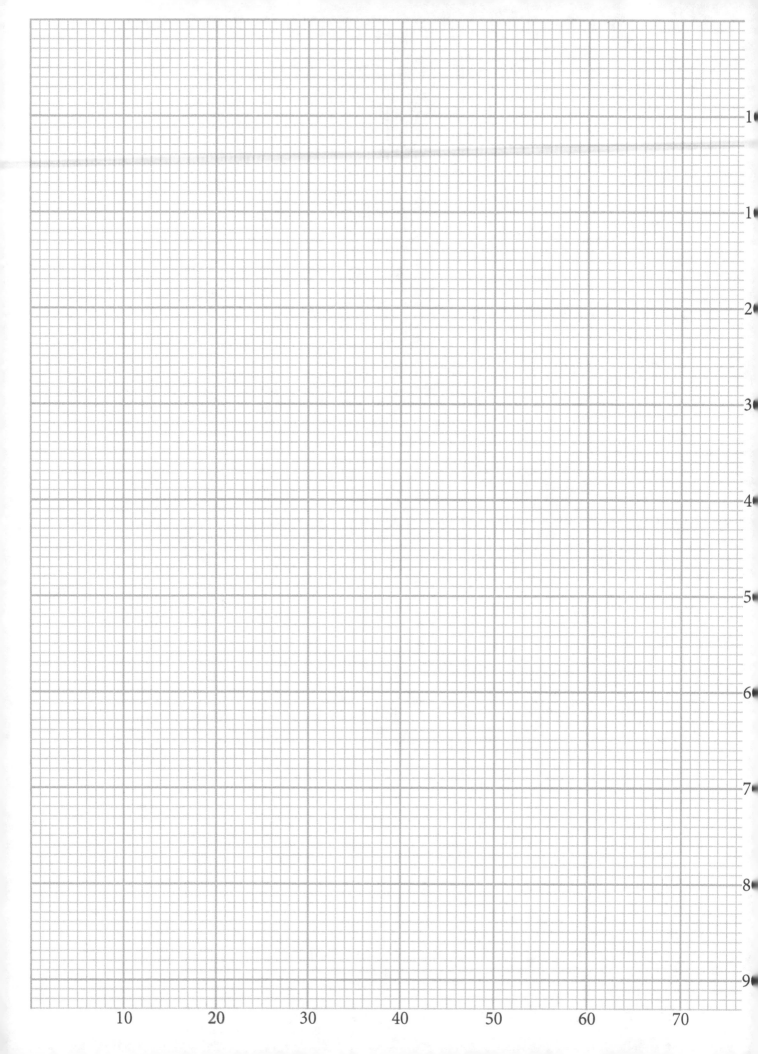

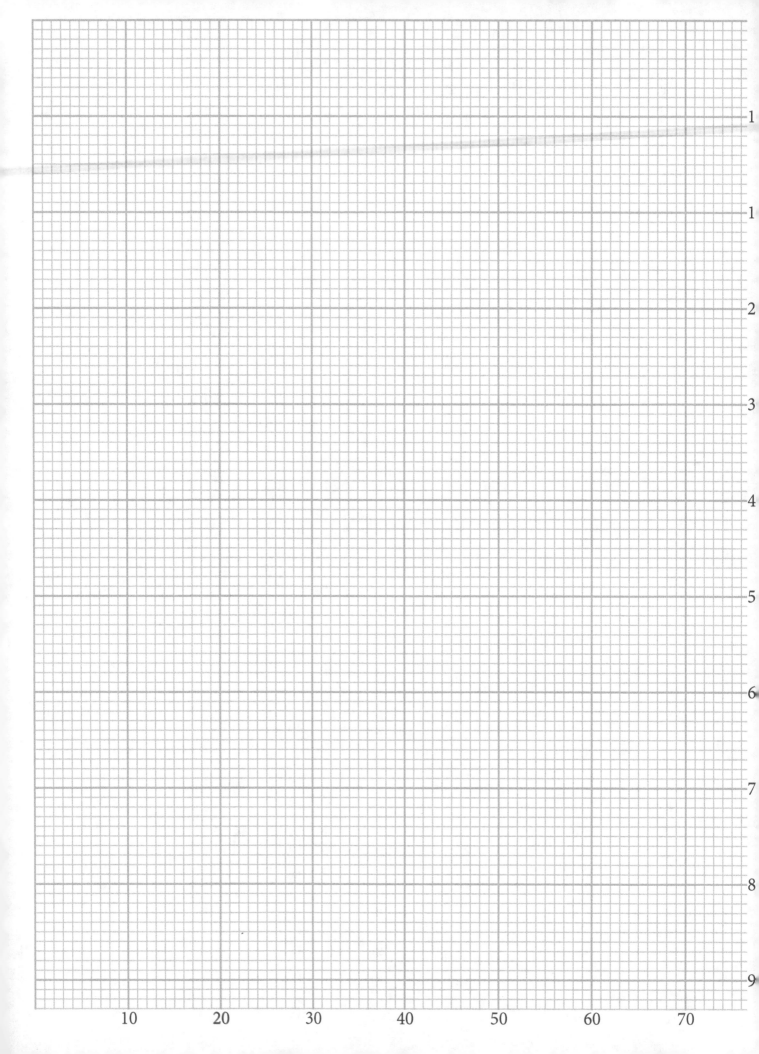

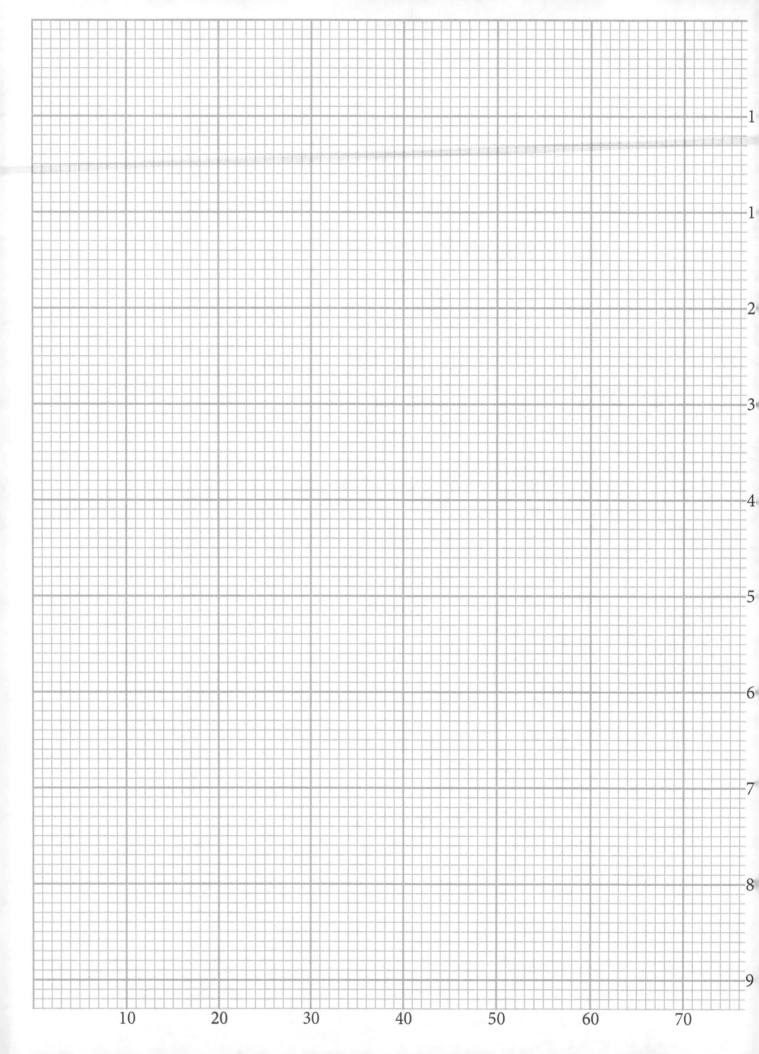

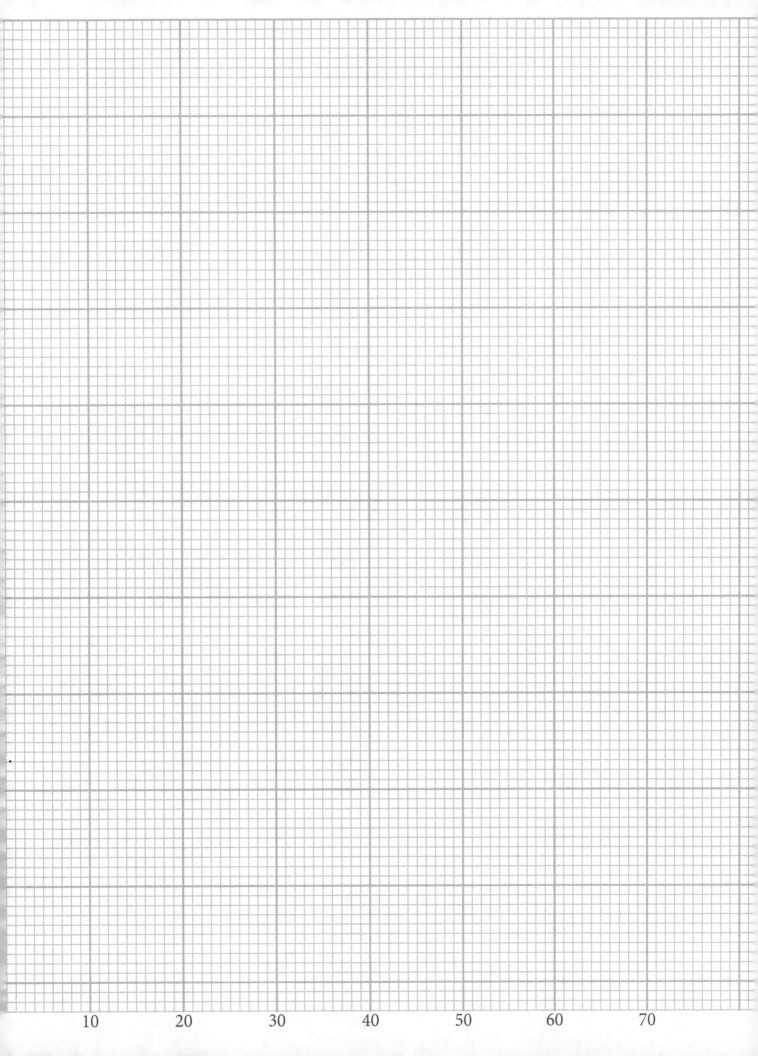

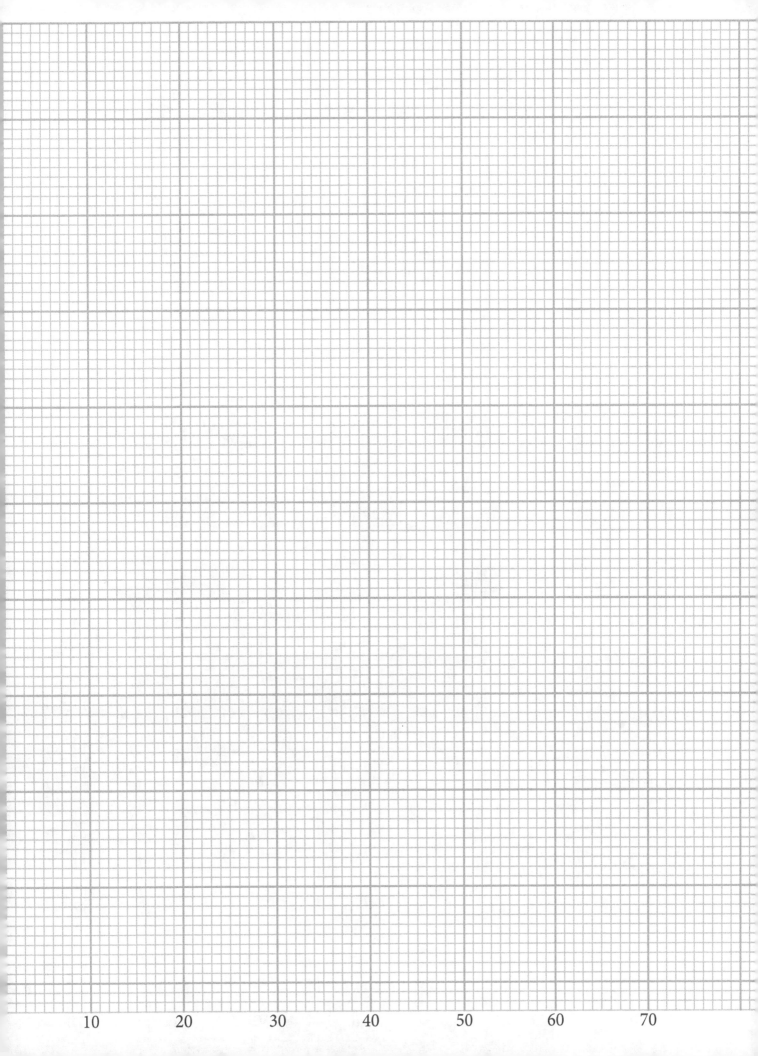

Made in United States
Troutdale, OR
11/28/2023

15092182R00060